This is You

DEPRESSION

Creating Your Path to Getting Better

FAITH G. HARPER, PhD, LPC-S, ACS, ACN

Microcosm Publishing
Portland, OR

This is Your Brain on

DEPRESSION

Creating Your Path to Getting Better

FAITH G. HARPER,
PhD, LPC-S, ACS, ACN

THIS IS YOUR BRAIN ON DEPRESSION

Creating Your Path to Getting Better

Part of the 5 Minute Therapy Series

First edition, first published 2016
Second edition, first published December 10, 2018

ISBN 978-1-62106-223-3
This is Microcosm #281
Cover illustration by Trista Vercher
Inside covers by Sophie Crumb
Book design by Joe Biel
For a catalog, write or visit:
Microcosm Publishing
2752 N Williams Ave.
Portland, OR 97227
www.microcosmpublishing.com

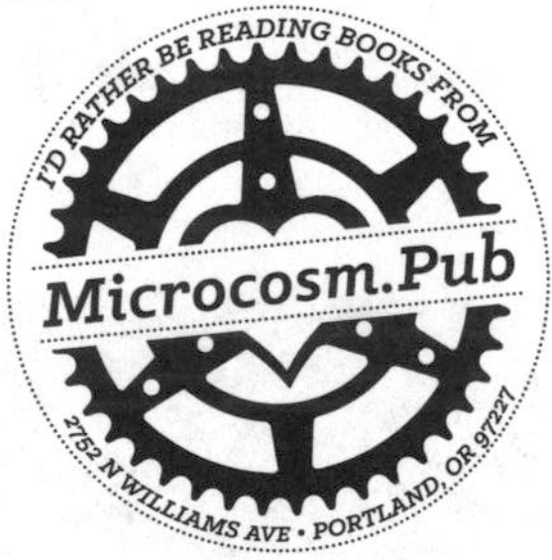

To join the ranks of high-class stores that feature Microcosm titles, talk to your local rep: In the U.S. **Como** (Atlantic), **Fujii** (Midwest), **Travelers West** (Pacific), **Turnaround** in Europe, and **Baker & Taylor Publisher Services** for other countries.

If you bought this on Amazon, I'm so sorry because you could have gotten it cheaper and supported a small, independent publisher at **Microcosm.Pub**

Global labor conditions are bad, and our roots in industrial Cleveland in the 70s and 80s made us appreciate the need to treat workers right. Therefore, our books are MADE IN THE USA and printed on post-consumer paper.

Library of Congress Cataloging-in-Publication Data
Names: Harper, Faith G., author.
Title: This is your brain on depression : finding your path to getting better / Faith G. Harper, PhD, LPC-S, ACS, ACN.
Description: Second edition. | Portland, Oregon : Microcosm Publishing, 2018. | Originally published: 2016. | Includes bibliographical references.
Identifiers: LCCN 2018024077 | ISBN 9781621062233 (pbk.)
Subjects: LCSH: Depression, Mental. | Depression, Mental--Treatment.
Classification: LCC RC537 .H347 2018 | DDC 616.85/27--dc23
LC record available at https://lccn.loc.gov/2018024077

Microcosm Publishing is Portland's most diversified publishing house and distributor with a focus on the colorful, authentic, and empowering. Our books and zines have put your power in your hands since 1996, equipping readers to make positive changes in your life and in the world around you. Microcosm emphasizes skill building, showing hidden histories, and fostering creativity through challenging conventional publishing wisdom. What was once a distro and record label was started by Joe Biel in his bedroom and has become among the oldest independent publishing houses in Portland, OR. In a world that has inched to the right for 80 years, we are carving out a place in the center with DIY skills, food, bicycling, gender, self-care, and social justice.

eating well
ONLY $2.99!
sleep + Rest
exercise
herbs, meds, etc
having a schedule
REFILL 3 weeks ago.
1: BAND PRACTICE
3: TRAPEZ LESSONS
7: EAT
8: SLEEP!
FISH OIL

CONTENTS

1. WHAT IS DEPRESSION?

epression is a MOTHERFUCKER.

Let's just put that shit right out there. And may it serve as fair warning for how I roll, language-wise, as an author: I think "fuck" is a perfectly cromulent word.

Depression is one of those words we throw around and use as a label so indiscriminately it's lost its meaning. I've been guilty of it. Maybe you have, too.

I used the word depressed to express how I felt when Whole Foods stopped carrying my favorite ginger cookies, even though *pissed as fuck with a preposterous sense of entitlement* would have been a way better description of my state of mind.

Check out your social media account and someone is stating they're depressed because they had Morrissey tickets and he canceled the show (and seriously, he *always* cancels the show; you are not only *not depressed*, you should also be *not surprised*).

Depression is not your team losing in overtime, losing your favorite watch, getting fired, or breaking up with a partner. Granted, all of these things have different levels of suckitude, but at their core they are all losses that cause understandable levels of grief. Grief and loss can absolutely be traumatic and can absolutely lead to depression. But with proper space and time to heal from grief, we heal. Depression is a far more insidious problem, and sometimes it doesn't have anything to do with an identifiable loss.

What is depression? Depression is ***a biochemical learned helplessness response*** to stress. Depression is the body's way of saying *nothing I do is going to help anyway, it all sucks ass no matter what.*

In this way, depression is a lot like anxiety. Anxiety is a biochemical over-response to stress hormones. It's the body trying to go into survival mode to protect itself, based on what it thinks to be true.

Robert Sapolsky, the brilliant stress researcher who wrote *Why Zebras Don't Get Ulcers*, defines depression as "a genetic-neurochemical disorder requiring a strong environmental trigger whose characteristic manifestation is an inability to appreciate sunsets."

I define it as ***a clinical case of the fuck-its.***

Depression is not the same thing as sadness, grief, coping with trauma, or coping with loss, although depression can have its onset in any of those things. Depression is the complete shutdown of all the things that make being human a joyful experience. Depression is the body saying *if nothing I do makes any difference, there is no point in enjoying ANYTHING.*

The biggest, most consistent symptom of depression is *anhedonia*, which is a tongue-twister way of saying *an inability to feel pleasure*. If you look at that word, you can see it essentially means not-hedonistic. If you struggle with depression, you have all kinds of feels. Guilt, shame, anger, irritability, hopelessness, overwhelming grief. But you rarely have experiences of pleasure, gratitude, connectedness, and joy. And if you do reach out for them, you feel them snatched away more often than not. Depression is the thief of all the wonderful things that make human-ing worth it.

The word depression comes from the Latin root *deprimere,* which means *to press down.* Yup, exactly. Depression operates like a French press . . . but instead of getting coffee out of the smashed-down grounds, you get a depleted and wrung-out human being.

2. WHAT TRIGGERS DEPRESSION?

Why do some people end up with depression? Fancy researcher types don't actually know. I mean, not really. We have a pretty good idea that there is ***a significant genetic predisposition to depression*** and are pretty certain that ***an environmental trigger is also required***.

This means if we are talking about the infamous "nature versus nurture" question, the answer is "yes, both." In order to develop depression, you will likely need both the brain wiring (because it runs in your family) *and* some fucked-up shit to trigger the disease actually kicking in. You know, like *trauma*. (I'm a trauma therapist; I'm always seeming to go there.)

So when we are talking about nature versus nurture, a better way to think of it is as *nurture informing nature*.

Very little of our genetic programming is set in stone. Only 2–5% of all diseases are related to a single faulty gene. Many, many, many diseases, however, are lurking in our DNA and can be turned on by the right conditions. The super fancy term for this is ***epigenetics***.

If depression runs in your family, it could be because of epigenomes that were turned on in your parents or grandparents by traumas they experienced. Then these genetic changes passed down to you. So, you were born with the wiring for depression. And if their history was bad enough, or enough bad things got piled onto you, then you are far, far more likely than other people to struggle with depression.

In a practical sense, this means your brain is wired for neurotransmitter misfiring. Lemme bust out the science for a minute.

So, nerve cells in the brain, called neurons, have to communicate with one another to accomplish any shit you gotta do in your pursuit of human-ing.

Don't be put off by the word "neuron" even if it sounds like I'm about to science you into a coma. Neuron truly only means "nerve cell." These transmit, receive, and process information throughout the body. *Think of neurons as a group of gossips that have some hot dirt they want to snark about.*

Neurons don't actually connect to one another in a physical, holding-hands type way—there are small pools of space in between. Instead, neurons have receptors on their ends that let them catch messages being sent out by other neurons. The receptors are tiny molecules chilling out and waiting to have a convo. These spaces between neurons are called synapses. *Consider the synapses to be like the texting app that the gossips are using, and the receptors are the phones themselves.*

The messages are sent through these spaces via neurotransmitters. Neurotransmitters are also tiny molecules: they catch hold of the molecules in the receiving neuron in a key-master-to-gatekeeper type love match to get a message through. *Neurotransmitters are the actual messages that the gossips are texting to each other.*

Like with any other convo, there has to be space and distance in order for messages to be received and processed properly. Receiving neurons need to rest and digest between messages. Unlike *actual* gossipy assholes, they aren't overly social motherfuckers, it appears.

During these resting periods, they dump the communication molecules back into the synapse, where they are supposed to be collected back up by the neuron that first sent the message. That's called *reuptake*, and it happens so the neurotransmitters can chatter again another day.

When things go wrong in a brain-fuckening way that leads to depression, the problem is somewhere in this communication process. The neurons aren't a good match in the convo. Or the rest time is disrupted. Or the neurotransmitters don't reuptake properly. We don't know. Anyone who tells you they know for sure what's going on there is a fucking liar. People way smarter than me have some intelligent theories, but *no one knows for sure.*

We do have a pretty good idea that depression is tied to three particular neurotransmitters: dopamine, serotonin, and norepinephrine. And possibly also glutamate. So there's that. But there are tons of "too much" versus "too little" debates. We do know that these are the chemicals that manage our sense of pleasure and our ability to cope with stress. So when they get jacked, our brain gets fucked. Depression ensues.

And when we start to feel like shit, we spiral down even more, which reinforces the disconnect in the neuron convos. So our brains start to think that nothing is controllable, fixable, or ever gonna be good again. I'm gonna refer to this tendency as *the depression funnel.*

3. THE DEPRESSION FUNNEL

We talked about depression as a clinical case of the fuck-its. But figuring out how to conceptualize that in a way that makes sense can be really difficult. It's kinda like the famous quote from the Supreme Court Justice Potter Stewart when discussing the nature of porn: *you know it when you see it*. Trying to separate porn from art can be as difficult as trying to separate out depression from grief or any number of human experiences. Then I happened upon the funnel analogy.

Marie Asberg from the Karolinska Institute in Stockholm, Sweden, specializes in burnout and created a way to conceptualize the burnout experience as an *exhaustion funnel*. Her essential idea is that we sort of slide down this funnel into a level of exhaustion that we know as burnout.

It's a bone-tired, over-everything, no-joy-in-the-things-that-matter level of exhaustion that a nap can't cure.

The slide down happens when we get busy and we give up the things we deem nonessential, like healthy food and body movement, even though those are the things that nourish us the most. Darren Littlejohn (author of the book *The 12-Step Buddhist*) uses the same analogy to demonstrate how a narrowing of perspective, support, and experience leads to relapse in addiction.

I *love* this idea and love the idea of using it to create a mechanism for understanding how depressive episodes are invoked. This image is my hack of her concept, considering the different way mood disorders can be experienced (rather than just the more generic idea of exhaustion) and the types of things that dump in, creating the problem to begin with.

The arrows at the top represent all the things that make us predisposed to depression from day zero and all the factors that make recurrences of depressive episodes more likely. They're all the life bullshit that gets dumped in every day. There is a ton of stuff that we have zero control over here: family history, trauma history, life bullshit (what researchers call significant life events, or SLEs), the age of initial onset of

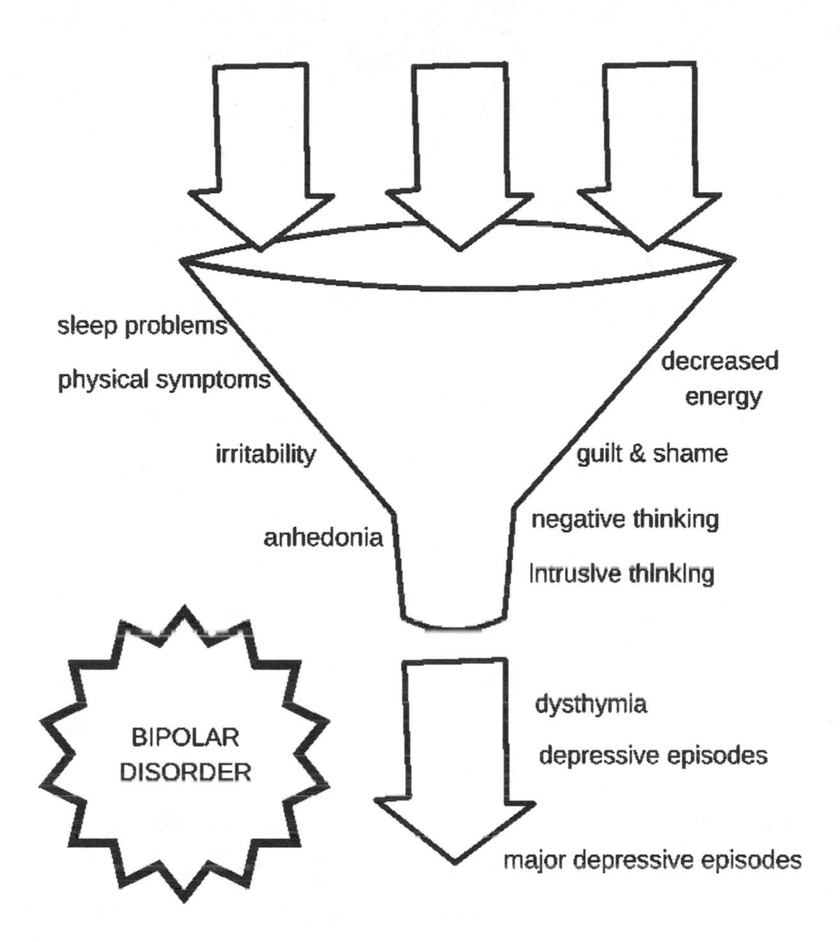

depression, or other diseases (either mental or physical) that can invoke *more* depression.

When we are at our healthiest, we have a larger experience of life. Like the top of the funnel, you can pour in these stressors and have room for them to exist without spillover.

Then depressive symptoms kick in (sleep problems, decreased energy, etc.), and this is where the funnel starts to narrow, because we start letting go of the things that seem the most optional. You know. The self-care stuff, like getting enough good quality sleep, eating the foods that are healthiest for our bodies, moving those same-said bodies around a little bit, etc.

The problem is, the optional things are the things that most nourish us. So, we funnel into a narrower and narrower way of living as we are depleted more and more. Eventually the only things left are the stressful things we should attend to, rather than the things that give us joy. *Trying to manage depression leads us to do the exact things that reinforce the depression instead.*

This whole funnel concept helps explain how depressive episodes can creep up on us. When we have all kinds of bullshit raining down, our default mode of managing these issues means we are sliding further down into the funnel. There is also no universal set point for realizing "oh shit, I need to get some help right now," so figuring out our unique funnel experiences and where the slide begins is an integral part of our individual recovery plans.

You'll also notice that bipolar disorder is sort of on this graphic, but sort of off to the side like it's its own little explode-y thing.

Because it totally is. Bipolar depressive episodes can operate in the same way, but manic episodes are the experience of an increasingly elevated mood (which can look like intense anger as much as it can extremely high energy). If the mania continues to spiral, it often leads to a pretty big kaboom, due to the behavioral choices that someone in that elevated state often ends up making.

4. WHAT KIND OF DEPRESSION DO I HAVE?

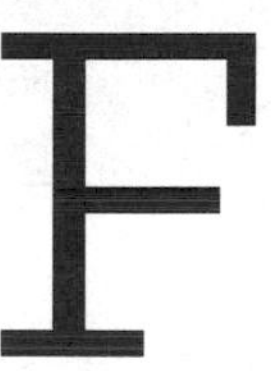

irst of all, I gotta say something here about diagnoses and the whole fucking point of them. And I'm saying this as someone who actively avoids diagnoses when at all possible.

All a diagnosis is intended to do is serve as shorthand communication between treatment providers. A diagnosis says *this person presents in this general cluster-of-symptoms framework*. So then providers can say, "Yeah, OK, got it. We've got treatment options that are pretty good at targeting that particular cluster."

And even then? There are enough cluster-of-symptoms categories that every one of us could have a diagnosis or ten if we really wanted to. The big difference between experiencing these types of things and needing treatment for depression (or anything else, as the case may be) is the impact that these symptoms are having on your life domains.

When I say life domains, I mean things like work, school, hobbies, sports, family functioning, romantic relationships, and friendships. Life domains are the things that are important about living. And something is a problem that may require a diagnosis if it interferes with some or all of the shit that makes life worth living.

There are some really great things about having a diagnosis. And some really horrible ones.

The big, great thing is the sense of relief. That something going on with you makes sense. It has a name. It exists as a real problem. It has treatment options.

The big, horrible thing is that we then so often become defined by this label. It becomes *who we are* instead of *something we have.* I've never heard anyone referred to as a cancerous person. But I've heard plenty of people referred to as a depressed person or a bipolar person. And that shit is NOT acceptable. This is a whole other conversation about the cultural norms around illness, so I'll tuck away my soapbox now. Because the language of diagnosis is the language we speak to access treatment.

So, without a total paradigm shift, it's the language we need to speak so we can communicate about care. And wellness. And healing. And reconnecting to the things that make life worth living.

But, I so want you to know that *you are not your diagnosis.*

There are things that are getting in the way of life being as great as it can possibly be, and maybe we can help diminish the negative impact they are having on your life.

Depression, Dysthymia, Bipolar Disorder, Oh My!

Depression falls under the category heading of "mood disorder."

Of course, there are a *lot* of fucking different diagnoses under that heading. There are plenty of people who work in mental health who don't fully understand all the differences, which means that the people they treat may not fully understand all the differences either. Part of my work as a therapist has been to help people just get a better understanding of their diagnosis and what it actually means. A diagnosis can provide a huge sense of relief, but that doesn't mean that you now magically understand what's going on.

If you don't have a diagnosis but have some suspicions, the information in this chapter can also be helpful in parsing out what you bring in when you have a convo with your doctor or therapist.

In terms of providing an official diagnosis, the book we (clinical people who are allowed to diagnose by licensure and training) most often work from is the DSM-V (Diagnostic and Statistical Manual) and ICD-10 (International Statistical Classification of Diseases and Related Health Problems), as well as some of the more legit websites like *WebMD.com* and *MayoClinic.org*.

So the information here is based on the information you would find in these places. It's not perfect (holy shit, far from it), but it's the shared dialogue that clinical people use. Which means, if we are operating within the traditional treatment system, it's the language we gotta speak.

Dysthymia (Persistent Depressive Disorder)

The big depression diagnosis we always think of is major depressive disorder (MDD), but there is an insidious little gremlin of a diagnosis called dysthymia (also known as persistent depressive disorder) that many people who don't qualify for the MDD diagnosis still experience. Instead of being mired in the black gunk, it's like perpetually wearing gloom-covered glasses. People with dysthymia are more likely to function better in daily life and seem more okay to outside observers than people with MDD. But they aren't really human-ing at a fundamental level. They can zip on their human skin and get through their day, but their insides feel jacked most of the time. Dysthymia can show up and couch surf on your ass for years at a time, with an intensity that can fluctuate quite a bit. If you have dysthymia, you may have periods of OK-ness, but those periods generally don't last for more than a month or two before the motherfucker is back on your couch, eating all your chips again.

Symptoms of dysthymia can include:

- *Less interest, or no interest, in daily activities (anhedonia, as mentioned above)*
- *Feeling sad, or down, or just kind of empty*
- *Feeling pretty hopeless about life*
- *Low energy, feeling low level tired all the time (whether getting sleep or not)*
- *Feeling like you can't do shit right, having lots of negative self-talk or low self-esteem that isn't really related to reality (because while saying you are bad at dunking when you're 4'11" is likely legit,*

saying you are fundamentally broken, shitty, and unlovable isn't)

- *Trouble paying attention, concentrating, or making decisions*
- *Anger or irritability*
- *Serious decrease in productivity of effective task completion*
- *Avoiding social situations and activities (the ones you would actually like to do in theory, or used to like to do)*
- *Worry, guilt, or shame*
- *Changes in eating (either overeating or not wanting to eat at all)*
- *Changes in sleep patterns (sleeping too much, not sleeping enough, sleeping badly)*
- *Stuck in the past and negative experiences that happened*

Major Depressive Disorder (MDD)

An actual diagnosis of major depressive disorder requires that anhedonia be present every day for at least two weeks. Other symptoms that are also really, really common are:

- *Loss of interest in all the things fun, excellent, and the point of being human (anhedonia, like I mentioned above)*
- *Low energy/fatigue*
- *Low-level chronic pain*
- *Headaches, stomach pain, or chest pain*
- *Jacked up concentration, difficulty making decisions*
- *Feeling guilty and/or worthless*
- *Sleeping a ton or sleeping for shit (not sleeping at all, or sleeping badly)*
- *Feeling either super restless or really slowed down (like moving underwater or brain wrapped in cotton)*
- *Intrusive thoughts of death (morbid ideation) or suicide (suicidal ideation)*
- *Change in eating habits (and 5% or more change in weight, either up or down, because of it)*
- *Irritability, anger, low distress tolerance*

I know, this feels like exactly the same list as dysthymia. And it essentially is, but it's really a matter of degree and the number of symptoms present at that degree. The essential question is: How well are you functioning and how much effort is it taking for you to do so? This doesn't necessarily mean that if you have dysthymia you make it through your day OK and with depression you don't. Everyone's experience is unique, and their symptom constellation is unique. I don't have a magical equation answer for you on this, so it may be something you have to unpack with your doc or therapist.

Bipolar Disorder

Then there is bipolar disorder (which used to be called manic-depression).

People who have bipolar disorder cycle through highs and lows. It's not depression and non-depression, but depression and mania, an intensely elevated mood. It's not necessarily fun and happy—it can also be high agitation, irritability, and anger. Unlike normal levels of elevated emotions, mania takes us over completely. Someone in a manic state really struggles to control their actions because their brains are in over-fire—no passing go, no collecting two hundred dollars.

Mania can cause other symptoms as well, but seven of the key signs of this phase of bipolar disorder are:

- *Racing thoughts*
- *Talking really fast*
- *Not needing much sleep to function*

- *Being easily distracted*
- *Feeling really restless*
- *Acting impulsively*
- *Being confident in your abilities far beyond your actual skills*
- *Elevated mood (either super high and happy or super angry and irritable)*
- *Making poor decisions/choices, engaging in risky behaviors (like with sex or money)*
- *Break from reality (psychosis)*

Some people with bipolar disorder may not show symptoms of full-blown mania, but they do swing into a pretty intense elevated mood. Not as extreme as it could be, but they're clearly not their normal selves. The clinical word for this is "hypomania", and is usually a warning sign that someone is in an upswing and is in need of help right away . . . or three days ago, if possible. Some people stay in hypomania and never get any worse (and really, hypomania is no picnic either), while others find that hypomania is their pit stop on the way to a full-blown manic episode.

To break it down a little bit more, there are two diagnoses possible here: bipolar I and bipolar II. People with ***bipolar I*** disorder have had at least one manic episode that was either followed by or preceded by either a hypomanic or depressive episode. Generally speaking, people with bipolar I experience depression symptoms four times *more* than they do mania symptoms.

People with ***bipolar II*** disorder have had at least one hypomanic episode along with a depressive episode but have never been in full-blown mania. Bipolar II is *not* a milder version of bipolar disorder, trust me. Mania can be severe AF, causing huge problems in school, work, and relationships. But people with bipolar II usually have longer and more severe depressive episodes, which can cause some huge life functioning problems as well. The four to one ratio I mentioned in the paragraph above for bipolar type I? In bipolar type II, people generally experience depression symptoms 35 times more than they do hypomania symptoms.

Other terms you may see associated with a diagnosis of bipolar disorder can include *mixed episodes* (depressive and manic symptoms happening at the same time, like one or the other wasn't difficult enough to handle) or *rapid cycling* (which means you move between depression and mania four or more times during a year's period).

Bipolar disorder can occur at any age, but it's most likely to first sprout up like a nasty little weed when we are in our teens or early twenties. It may not get diagnosed until much later, obvs, but the signs and symptoms are usually there at some point during puberty.

Why am I talking about other mood disorders in a book about depression? Bipolar disorder has a huge depression component (no shit) but is not the same disease. I'm including this info because the stats on people with bipolar disorder being misdiagnosed scare the fuck out of me. Way too many people who have bipolar disorder are misdiagnosed as having "just" depression. Up to 20% of people who go to the doctor because they think they have major depression actually have bipolar disorder. Especially if you have bipolar type II, the hypomania symptoms may not be noticed, or they are misjudged as being relief from the depression. So bipolar disorder needs to be part of *any* mood disorder conversation we're having.

On average it takes a good *decade* to get real treatment for bipolar disorder. This is caused in part by the struggles with getting an accurate diagnosis; at least half of people with bipolar disorder have seen *three* damn professionals before getting a correct diagnosis. The chance of having another diagnosis along with bipolar disorder (like substance abuse or anxiety) can also make diagnosing legit difficult.

Anyone who is being evaluated for depression (you or someone you are dragging in for an appointment) should be asked about their whole lifetime history of symptoms, really looking at any past manic or hypomanic episodes. And anyone put on antidepressants should be watched carefully. Not only is there a spike in suicidal ideation, which is common enough for anyone starting an antidepressant (the famous black box warning), the meds can also invoke the person's first manic episode.

However, this book is by no means useless if you're bipolar. Whether type I or type II, you are generally spending the majority of your time combating depression. The info in this book about fighting that hell-beast still applies.

Other Possible Diagnoses

If you are thinking you or someone you know might have a mood disorder, there are some other potential causes to consider.

Drugs and alcohol are the big and obvious ones (e.g., being on drugs can look like a manic episode and coming off of them can look like a depressive episode). But more than that, the relationship between addiction and depression is a damn strong one. Depending on your ethnicity, the chance of you having both a mood disorder and a substance use disorder is at *least* 50%. It can be as high as 72%. And which comes first? Depends on the person, right? Some people end up using drugs and alcohol to medicate their mood disorder. Some people use drugs and alcohol for different reasons and end up developing a mood disorder because the drugs and alcohol change their brain wiring. It used to be that most providers wouldn't treat the mood disorder unless you got sober first. Fortunately, now we realize that's entirely unhelpful to the process. Though, getting the right treatment and the right medication can be difficult if we don't know what is caused by drugs and alcohol and what is just brain chemistry. Most prescribers will, understandably, limit what they prescribe if it will interact badly with drugs and alcohol as well. So tell them the truth. It's far better than ending up in a coma.

Anxiety disorders are not considered mood disorders (they are classified as their own category) but they pair so frequently with mood disorders we should just call them the Heathers and be done with it. It makes sense that mood disorders and anxiety disorders would go hand in hand when you think about it. Your stress system is all kinds of misfiring, so you go into overdrive . . . which we call anxiety. If this becomes status quo, your body eventually gets worn the fuck out. Your cortisol levels start to drop, and that leads to depression. Some people notice that after years of chronic anxiety they start to experience depression. Some people go back and forth between the two or experience both concomitantly because their HPA axis system (that's the hypothalamic-pituitary-adrenal axis if you're nasty) is in constant flux of trying to respond, shutting down in exhaustion, then trying again to respond. I do talk more about this in my book about anxiety if that's something you experience consistently.

Postpartum Depression—Research shows that one in eight moms (and other individuals with a uterus who have carried and birthed a child) struggle with postpartum depression in those first months after giving birth. Why so high? First, there is a huge drop in estrogen and progesterone in the body after giving birth, which can trigger the changes in how chemicals fire in the brain. Also, there is the straight-up lack of sleep that all parenting people with a new baby experience. And if you are already exhausted (because you just birthed a human-fucking-being out of your body!), you aren't getting the recovery time you really needed. Treatment for postpartum depression will be very similar to that of any other version of depression (medications, talk therapy) but may not be a forever thing, the way treatment for major depressive disorder may be. Also? Sleep when the baby sleeps. Fuck doing the dishes. Get paper plates if you need to. Take care of you so you can take care of your baby. Postpartum depression was so much me, y'all. And I learned better self-care skills after having my daughter, so it was far more manageable when I had my son.

Medical conditions like multiple sclerosis, a stroke, and Cushing's disease (to name a few common culprits) can either look like a mood disorder or make the symptoms of a mood disorder worse. Pregnancy can also change your symptomology, as can just the plain ol' weather (seasonal affective disorder is a real thing, y'all). Medical conditions can definitely make depression worse, but certain medical conditions can mimic depression. And if a doctor is throwing an antidepressant at you, they might be treating the wrong thing. These are things that may need to be investigated when you are looking for the right treatment options (which are generally developed by hitting on the right diagnosis). Doctors need to know ALL your symptoms, even if you think they are dumb and unrelated. And all your family history, again, even if it seems dumb and unrelated. If initial treatments fail you, continue to fight to see specialists and screen other things out.

Example: I have a client who was diagnosed with bipolar disorder and put on mood stabilizers, but I noticed he had a lot of other weird prescription meds listed for blood pressure and diabetes on his intake form. I asked to see his recent bloodwork and had him fill out some assessments related to physical symptoms. (I should add I have a post-doc in clinical nutrition, but most therapists aren't also nutritionists.) He had some off-the-chain shit going on in his body. Severe vitamin deficiency, trouble processing fats, blood sugar stress, and chronic inflammation. He wasn't bipolar, y'all, he was irritable and irrational because his body was completely depleted. We got his doctors to adjust his medications, added some whole food supplements to help promote healing, and the difference was almost immediate.

One last option to consider? ***Cyclothymic disorder.*** This diagnosis is related to periods of hypomania followed by periods of dysthymia (depression that is sub-threshold to major depressive episodes). Now, we've probably all qualified for that diagnosis in some point in our lives, so this diagnosis is more about *time*. This isn't about the sort of mood swings we all suffer when shit's fucked. *It's pervasive over time, regardless of circumstances.* If you are an adult getting diagnosed, this has been going on for at least two years. If you're a teen or a kiddo, then at least one year.

A lot of researchers think it's a milder form of bipolar disorder (since it typically appears in people who have family members with bipolar disorder). But milder doesn't mean unserious. Cyclothymic disorder is far more chronic and pervasive than bipolar disorder, and most people don't seek treatment because they manage through just enough. The problem with that is, your chances of developing bipolar disorder are up to 50%. So managing through cyclothymia without help makes as much sense as changing deck chairs on the Titanic. Auntie Faith hates to worry about you, so get treatment if you need it, OK?

There are a lot of other possible diagnoses that we use when we aren't sure what's going on, ***mood disorder NOS (not otherwise specified)*** being the big one. This means that we know there is a disruption in the brain chemical balance that is affecting mood, but the symptoms we see don't really match a standard diagnosis. NOS diagnoses are used when the diagnosis manual fails to serve us as intended.

No matter what the actual fancy label, these are all just a way of communicating that the brain is misfiring the chemicals that cause the feels. And the feels are what make life a win. Depression and other mood-based diagnoses are some seriously soul-suffocating awfulness: real-life

Dementors. J.K. Rowling has struggled with depression; she knew exactly what she was writing about.

Down in the appendices are some stats about depression and the impact it has on not just the people who struggle with it, but society in general. It isn't necessary information for your getting better part, but it can be useful for context about the scope of the problem. Or if you gotta school some toolbox who doesn't think depression is that big of a deal.

5. THE GETTING BETTER PART

The bad news is there is no magical path for healing depression. However, that's also the good news. That means you get to find the path that works best for you. And fuck anyone who tells you you aren't healing correctly.

Because there is no magical answer about what treatments you should seek out, the important thing is to be aware of the many options available for you to choose from . . . especially when there are people who are going to try to push their worldview about treatment on you.

Maybe the hardest thing to do when struggling with depression is to actively seek treatment: researching options, advocating for yourself, following through on treatment, and finding new and better options when your first (or seventh) attempt fails.

This is because depression, along with being a motherfucker, *lies* like a politician trying to convince you that drilling for oil in a national park is a good idea.

Remember all the stuff about depression being a biochemical learned helplessness response? This means that depression by *its very definition is* a piece of shit who is going to try to gaslight you into thinking that (1) everything is hopeless and (2) treatment is a waste of time because nothing will work.

If we aren't supposed to buy that bullshit from a toxic ex, we sure as hell aren't going to let our own brains convince us we aren't worth the effort.

Because here is the good news: all that stuff about epigenetics and certain switches turning on for depression? If we can turn them on, then we might be able to turn them back off again. At the very least, we know how to not just target what's going on with the neurotransmitters, we know how to watch for and fight the toxic funnel effect.

Complete and permanent remission? Possible. And I've seen it happen. But even if it doesn't? Depression can be tamed, leashed, and trained to not dominate every waking moment of your existence. *Just like any other chronic disease.*

Whoa, lady, does this mean I may not to be on meds forever? Maybe I won't pass it down to my kids? Maybe it won't keep getting worse year after year like it has been?

More hard maybes there. Blech. I wish I knew.

Depression and other mood disorders are really damn difficult to treat. Research demonstrates that up to 39% of the people diagnosed with depression will still meet criteria (meaning still be clinically depressed) a year later. And even if effectively treated, your chances of having another depressive episode are still 26% higher than someone who hasn't suffered from one. A lot of this has to do with how hard depression is to treat. And, honestly, we are just starting out how to treat it effectively because we are treating the underlying causes, not just the symptoms.

Essentially, once our brains make a connection between sad emotions and negative thoughts, that becomes a groove of thinking and reacting. So even "normal" life sadness starts connecting to the types of toxic thoughts that are connected to major depression. There doesn't have to be a huge traumatic loss to start back down into the depression funnel again. Which means it can be really fucking difficult to pinpoint the trigger. Because honestly? The brain's own neg-gremlins are the damn trigger.

I can tell you the people who come to therapy have a way better handle on their mood disorders if we unpack the trauma fuckery. They manage present and future triggers way better. Sometimes they aren't nearly as impacted. At the very least they go, "Shit just got real again, I need some Buddhadamn help right about now." If they are on meds, they are often able to at least decrease, or find ways of not having to increase year in, year out like they had been. And yes, I have seen complete remission of symptoms a number of times. It is possible.

Allopathic and Naturopathic Meds

One thing I really wanted to do with this book is talk about the vast variety of treatment options out there, from the perspective of a "yes, and" therapist. Meaning I'm trained in the traditional medical model *and* in many of the complementary treatments that are gaining traction as evidence-based practice even though they aren't yet mainstream treatments.

Consider this section me trying to take at least a little bit of the overwhelm out of the getting better part of a depression diagnosis. Some of this info may be old hat. Some of it may be stuff you've never heard of before. It may be *"hell naw,"* or it may be *"interesting, and maybe worth a try."*

But it's all tied to the brain science info that I already ladysplained to you, and it's stuff that interconnects with other treatment options. So let's look at what's out there, yeah?

Allopathic meds refers to Western medicine: identifying and treating disease using specific remedies. Generally, by this we mean the prescription stuff. Or the over-the-counter stuff we grab at the store.

Naturopathic meds refer to more traditional treatments (herbs, minerals, nutritional supplements). And I'm saying "traditional" in terms of human history. Herbs and foods and the like are what human beings have used to treat illness for most of our time hopping around this planet. Allopathic meds are far newer forms of treatment for the human body. My grandfather-in-law is literally 102 years old and lived through this shift in thinking and treatment.

And now the big question is: What is my official recommendation? Am I pro-prescription or not?

I am pro-whatever-helps-people-get-better. We have enough stigma in our culture about mental health treatment, for fuck's sake.

If I see one more patchouli-infused, non-licensed "lifestyle guru" post one more meme on social media about how you don't need an antidepressant, you just need to walk in the woods (or hang upside-down in an inversion chair, or do a raw food cleanse), I'm going to LOSE MY SHIT.

Actually. Writing this book is my losing-my-shit response to treatment-shamers. Fuck all of them.

Prescription medications can play a vital role in your treatment. In my book *Unfuck Your Brain* (Microcosm Publishing, 2017), I used the analogy my friend Aaron Sapp, MD, let me steal. He explains the trauma reaction and the resultant brain chemistry changes and how medications can help. It's great, so I'm reprinting it here:

> *Imagine you are an Air Force base. Everything is fine, then all of a sudden all the lights go out and the radar goes dead. You aren't going to assume that just because a second ago everything was fine it still is. You are going to assume an attack. When you have a mood disorder, your radar has lost communications to the rest of the base, so it assumes an attack all the time. We are going to reconnect communications so your threat detector, which we assume is doing its job, is talking with the threat response unit again.*

Medications can absolutely save lives. They have done so, and they will continue to do so. However, they are not always a singular cure-all. And I get just as mad at singular-line-of-defense over-prescribers.

Because then we are ONLY medicating symptoms instead of the root causes. Anxious and depressed? We have pills for that. Then you go home with your pills and that's the only support you get until you go back to see the doctor for your follow-up appointment and fall into a routine of constant medication adjustment with little other support.

This leads to over-medication, tons of side effects, and then more medications to manage the side effects. We are seeing more and more stories of people medicated to the point of toxicity. And I realize that this is a whole other book, but many of the people being over-medicated are in the prison system. Or children in the foster-care system. And it would be the understatement of the decade for me to say how fucked up that is.

Prescription medications have their limitations. They don't always work as well as their manufacturers spend advertising billions for us to believe. Most people end up not taking them after a while for just that reason.

And even more to the point? The World Health Organization confirmed a long-term study showing that in third world countries where antipsychotic medications are not even available, recovery rates were actually *higher*. Because if the medications weren't available, they couldn't be the focus of treatment. So the causes were treated, giving people a sense of meaning and community. *And then they got better.*

Medications help adjust brain chemistry and manage the symptoms that we are struggling with. They give us the life jacket we need to stay in the game and build other coping skills and mechanisms of healing.

The more we can do to encourage the body's own ability to adapt and heal, the better. Medications can be an integral part of that journey—although they are rarely the only tool we use. Educating yourself and advocating for yourself about prescription medications will greatly increase the chances that they are used properly with you. Go check out the appendix of this book for a breakdown of different types of antidepressants and how to have a convo with your prescriber about the best options for you.

The point of treatment is to give the body what it needs to regain its own equilibrium and heal itself. This is where naturopathic medicine comes back into the conversation.

Naturopathic meds are generally broken into two categories: whole food supplements and herbal supplements. I'm a huge fan of using both in support of (and *sometimes* instead of) prescription medications.

Whole food supplements are literally *just food.* You're piling on the nutrition you need for healing. In a time period where we are all missing things from our diet that our ancestors got (e.g. depleted soil means we may not be getting enough trace minerals in our diet from the veggies we eat, even if they are local and organically raised), adding those things back in can be incredibly helpful for our bodies. My motto is you have to either eat it or take it. And honestly, getting enough of the nutrients I need through diet alone is nearly impossible. Whole food supplements can help bridge that gap without the side effects of the synthetic isolates found in allopathic meds and even most over-the-counter vitamins.

Herbal supplements work far more like medications do but also more like whole food supplements than you would first guess. Instead of treating a symptom, they are also operating to promote the body's own healing and ability to balance itself. That's why so many of the herbal remedies that are most effective for mood disorders are adaptogenic ones. This means instead of an upper or downer, they function as allarounders. They're like the thermostat on the wall that is going to kick in heat when it's cold or the a/c when it's hot as balls. Meaning they work by promoting the *stabilization of physiological processes.* I included some information in the appendix specifically about some of the more common naturopathic options for treating depression, including which ones are adaptogenic. You know, if I piqued your interest any.

Yeah, you might say, *but . . . aren't supplements complete bullshit? I've read the articles.*

Part of the reason that dietary supplements get a bad rap is because many of the ones on the market are complete crap. The New York Attorney General tested and sent a multitude of cease and desist letters to herbal supplement companies based on the fact that much of what they tested had *no active ingredient.*

And also? University of Guelph in Canada studied a bunch of supplements and found many unlisted ingredients within them, including things that could encourage an allergic response in someone taking them. And synthetic versions of the product, rather than the actual extracted herb or whole food, are generally going to have more side effects because the human body struggles to recognize them for what they are.

So we read about the amazingness of using something like St. John's wort, then we feel stupid and/or ripped off when it doesn't work for us. I had that experience with a cheap kava I tried years ago. It made me seriously irritable and more than a little batshit. I was afraid to try kava again until I learned more about finding and using quality products. I was great about educating myself on prescription meds, but it somehow never occurred to me that I should treat supplements just as seriously. So, takeaway? If you are using supplements (either herbal or whole food), work with a provider who knows their shit (an herbalist, Chinese medicine practitioner, clinical nutrition professional, etc.) and research anything you choose to take and the quality of the brand you are using, just like you would any other treatment.

And because that can be confusing as fuck, information on finding someone legit is included in the appendix under "Finding Treatment Providers."

Traditional Talk Therapy

So, yeah. This is my jam. I'm a licensed professional counselor and a talk therapist through and through. Talk therapy has a great capacity to heal, in support of other treatments or sometimes alone. A good therapist has the benefit of their training and a perspective on your life that you don't have because they aren't living the experience you are living. They can provide insight, coaching, and interventions to help on your getting-bettering journey.

Some people find that talk therapy gives them the support they need to manage their depression and others use talk therapy in conjunction with medications. I have a friend, a nurse practitioner, who flat out refuses to see people who aren't also getting some other form of therapeutic support. Meds are generally just not enough. Research does show that the two together are far more effective than medications alone. I have seen a lot of people be able to greatly diminish their use of medications, or not need them at all, by doing some seriously tough work in therapy.

And, yes, trauma work can be a huge part of that. This may not necessarily mean talking about the trauma but always means learning skills to deal with the triggers. If you know, or at least have a hardcore suspicion, that your mood disorder has a big-assed trauma taproot, then it might make a fuck-ton of sense to treat the trauma along with the other symptoms.

Other Complementary Therapies

So here is where we talk about the treatments that our ancestors used and used effectively that were then tossed aside when modern medicine came along with its pills and surgeries. What used to be our first course of treatment now has gotten a bad rap as woo-woo bullshit. And it's entirely legit to not want to spend your hard-earned money and even harder-earned free time fucking around with woo-woo bullshit. But a lot of the things that have gotten a bad rap actually have a lot of research backing up their efficacy, even if, because they don't benefit the pharmaceutical industry, they are generally discredited and not covered by insurance.

And many of them can be used either alone or with Western practices (like traditional talk therapy or allopathic meds). Like naturopathic treatments, these are things that are designed to encourage the body's own capacity to heal and self-regulate—to keep the funnel wide enough to manage life stressors without collapsing in with an ever-narrowing perspective and ability to manage life.

One of the biggest supporters of complementary therapies (at least in my hood) is the Veterans Administration. We have a program in town that provides massage, acupuncture, reiki, and EFT (more on all of those below) to local veterans.

Our local VA has noticed that the people who do that work also do far better with their VA providers in therapy, med management, and symptom reduction. They actively recommend complementary therapies *and* have been training their own people in some of these therapies. As an example, some of their MDs are now also using acupuncture in the clinic.

Y'all. If the military-industrial complex thinks reiki works, it might just mean that reiki fucking *works*. So here is my list of complementary treatments that I encourage, have experience with, and have found the research behind. It's not a complete list (and I would actually love to hear if you have any additions you think I should consider).

Acupuncture/Acupressure/EFT

Acupressure and acupuncture use the same principles, but ***acupuncture*** involves using the actual needles in the skin while ***acupressure*** involves the tapping of certain points instead of breaking the skin.

However, whether tapping or using needles, it works by stimulating certain points on the body to promote healing and/or reduce pain. What is really interesting is that as we learn more about the vagus nerve system, we are seeing lots of commonality in modern nerve mapping and five-thousand-year-old acupuncture charts. The vagus nerve is large and complex. Though it's a cranial nerve, it wanders down throughout the body to many other organs, sending a messages back to the brain. Recent research (led by Stephen Porges) demonstrates that the vagus nerve plays a huge role in managing our social behaviors and responses to stress and trauma. Mood disorders (like other mental health issues) have a whole-body response to what is being fired off in the brain. Acupuncture and acupressure meridians follow both vagus nerve maps and then even deeper through the fascia that runs throughout our bodies.

If you are interested in a combo deal of acupressure with talk therapy, some therapists use Emotional Freedom Technique (EFT), which includes acupressure and self-talk strategies. The EFT is something you do yourself, with guidance from the practitioner, using the same main activation points an acupuncturist would (bonus if people touching you squicks you out). The self-talk helps you reframe the stories your brain has been telling you while creating new ones in the process. There are tons of free videos that walk you through the basic process, though a therapist will help you modify the scripts to work through your specific situation.

Biofeedback/Neurofeedback/Alpha Stim Treatment

Biofeedback is the electronic monitoring of all bodily functions, which helps people learn to control responses that were previously automatic. ***Neurofeedback*** focuses specifically on the brain signals with the same intent: to help individuals learn to manage their brain responses.

We have far more control over our body and brain's responses than we realize, and both bio and neuro can be great ways to augment or even speed up our unfuckening by giving us immediate feedback when our brain and body start to get into fight, flight, and freeze mode. You essentially play a video game with your brain. It sounds as Tron as all fuck, but you are set up with a Pac-Man type game, or something similar, which you can only complete when you keep your brain waves in the optimal zone for your wellness. We know that so much of depression is caused by brain misfiring. Neurofeedback can help you intentionally reshape how the brain fires messages to mitigate this response.

I also include ***Alpha-Stim*** treatment in this section, since even though it is a passive treatment, it falls under the same principles. Alpha-Stims are designed to increase alpha brain waves (which are the great combination of calm and alert that we all crave). It works like neurofeedback except the machine does the work for you rather than you training that brain state yourself. You hook up the machine to your earlobes and turn it on, and it does the brain wave-changing work for you. Alpha-Stims have lots of research demonstrating that they help with sleep, pain, anxiety, and a host of other conditions. I do use an Alpha-Stim in my practice, especially when clients are working through a trauma narrative that is important to them but causing a lot of pain. The Alpha-Stim can really help them recover faster from one of those tough therapy days! I have also had clients use them to improve symptoms in their daily lives without other medications. Alpha-Stims have to be purchased with a prescription in the U.S., but any counselor can prescribe them like I do, not just a medical doctor.

Chiropractic Treatment

What? Chiropractic care for mental health issues? Isn't that for bad backs? Beyond, again, the fact that depression can manifest as physical pain, chiropractic is a holistic form of treatment that operates from the idea that adjustments of the spine and body can facilitate nervous system support. Pain and nervous system support? Totally huge parts of a trauma reaction for many people, which can be the genesis of depression. And sometimes these physical symptoms are far worse than the emotional ones. Many chiropractors (as well as massage therapists and acupuncturists) build nutritional assistance into their work as well.

Diet/Nutrition Changes

When we are stressed, we crave sugar like whoa. The sugar may be the only high you get when your depression is bad. The brain also needs glucose to maintain willpower and energy . . . which is why dieting is so hard. You are deprived of the glucose you need for willpower. Typically, the more stressed and busy we are, the worse we eat. So it's a vicious fucking cycle and ridiculously frustrating.

I know there are lots of nutrition wars out there and figuring out what is the best plan for you can be exhausting (paleo? Vegan? Gluten-free? *The fuck am I supposed to eat*?).

Short answer? Our body works best when we take care of it, eating the whole and healthy foods humans ate for centuries. And any diet you follow is going to make you more mindful of what you are putting in your mouth, for damn sure. So I'm not hell-bent on you choosing one over the other. In fact, everyone has differing dietary needs, which is why my cookbook (*The Revolution Will Include Cookies*, Say Something Real Press, LLC, 2016) offers variations in each recipe for some of the more popular ones.

And seriously, getting help from a clinical nutritionist, Chinese medicine practitioner, or trained naturopath who incorporates nutritional work can be well worth the investment. I do clinical nutrition work in my practice and have worked with many people once or twice on diet modifications and supplements, and that's all they really needed: some basic assessments and advice to help them through the overwhelming information out there.

Nutrition and mental health really is a whole other book (gimme time!), but there are certain basics that can help enormously without getting into weird food cult status that we can cover without going down a huge rabbit hole:

- If we eat healthy about 85% of the time and enjoy treats about 15% of the time, we can maintain good functioning.
- Stay away from industrial foods as much as possible. The most important thing to remember about food labels? Trying to avoid foods that have labels. The more refined and machined a food is, the more likely it is for your body to not recognize it.
- The movement away from gluten is more about the genetic modification of the wheat in the U.S.

and less about gluten itself, at least for most people (celiac and people with severe allergies aside). While moving to France or Italy would be ideal, the best thing we can do in the U.S. is move away from gluten and genetically modified grains as much as possible. (I live in South Texas, so we live on corn tortillas anyway.)

- Many individuals that are entirely grain-free use coconut or almond flour. If you miss wheat like a physical pain, try original einkorn wheat flour in your baking instead of the crap from the grocery store shelves.
- Many people who can't tolerate dairy do fine with raw milks.
- Chemical sweeteners are total ass to your body.
- If you suspect something is making you feel worse, try dumping it out for 21 days. See how you feel. Add it back in. Notice a difference? Your body will totally tell you what it needs.

Energy Healing (Reflexology/Reiki)

Energy healing is one of those things that seemed super weird, even for me, for many years. Then I read more about it, and tried it for myself, and WOW.

So, ***energy healing*** is based on the idea that our bodies operate on all these frequencies that we can tap into to promote our healing. Weird? Not so much. One study showed energy healing being as effective as physical therapy. UCLA now has a whole fucking *lab* that studies electrical activity in the body. And UCLA is a state-funded school. Tax money invested in energy healing: that's some serious street cred.

Reflexology focuses on applying pressure to areas of the ears, hands, and feet, with the idea that these areas are connected to other points throughout the body (and polyvagal theory bears this out). ***Reiki*** (a Japanese term for guided life force energy) is the channeling of energy from a practitioner (or from one's own self) into the person who needs healing in order to activate the body's own healing process. These forms of energy healing (among others) help us find the stuck points in the body where we tend to hold our trauma in order to better release them.

And BTW? Acupressure (tapping work like the EFT I talked about) is considered a form of energy healing as well as a variation on acupuncture.

Massage

Everyone knows what massage is, I don't have to explain it, but people are surprised when I suggest it as healing for emotional issues, not just physical pain. First of all, physical pain can absolutely be a symptom of depression. But even if that isn't going on, massage can be a safe way for people to learn to relax and feel comfortable in their skin.

So many times after a trauma we feel disconnected from our bodies. We hold so much of what we term "mental health issues" in our physical bodies. The point of massage is the ability to ground your self back into your body. This helps us recognize and catch our symptoms earlier on in the depression funnel and makes it far easier to prevent a full relapse.

I realize massage can be very triggering for certain types of trauma. Definitely don't force yourself out of your comfort zone. Some people are way more comfortable with a pedicure and a foot massage than with a full body massage. Some people prefer a hot bath or hot tub soak rather than having someone's hands on their skin. Anything that feels safe for you while helping you reconnect with your physical body can really help you manage your depression, not just the pain associated with it.

Light Therapy

Light therapy is simply trying to mimic the light of the sun using a special kind of lamp. You may have heard about light therapy being used for depression, specifically for seasonal affective disorder (winter blues thought to be caused by lack of sunlight). When my brother left sunny Texas to go away to college in Boston, his SAD was horrible. He went from thinking snow was pretty and exotic to thinking snow was some kind of evil plan from the universe to repeatedly beat him in the face. The light box our mom sent him was stupendously helpful.

There is a ton of research behind light therapy. One study found it to be just as effective for SAD as Prozac (minus the side effects).

Light therapy may not have a magical curative effect on more severe forms of depression, but it can absolutely help, working in conjunction with other therapies. There is a specific light spectrum that helps pain (red light), and there is a specific light spectrum that helps depression (blue light).

How so?

You may remember learning about rods and cones in the eye and how they generate our ability to see. But there are also the retinal ganglion cells (ipRGCs) in the eye that relay information directly to the hypothalamus instead of to the vision centers of the brain. Blue light tells the brain "*wakey wakey*," which ups our energy for daily tasks. (And as an aside, this is why having tech equipment on nearby when you are trying to sleep can fuck up your sleeping. The brain is all "*wakey wakey?*" and keeps fighting with your tired body to get up and do shit.)

I keep a blue light in my office that clients can use for a boost during sessions. You can get decent light boxes now for under 100 dollars if you don't mind doing a little research and price comparison online. A lot of people find it helpful. Others have reported that it's an irritant, so see if you can borrow one or try one out somewhere before you invest in your own.

Weighted Blankets

You may have heard of weighted blankets for helping to manage anxiety or soothe an individual on the autism spectrum. They also really support sleep and depression recovery. The pressure of the blanket engages the nervous system and helps the body to re-regulate in general. Weighted blankets are usually made with plastic pellets (which makes

the blankets machine washable and not as hot as, say, a heavy wool blanket).

In terms of weight, for adults, look for a blanket that is about 10% of your ideal body weight, and for kids and teens it will be about 10% of their weight plus a pound or two. An occupational therapist can help you figure out exactly the right formula for you if you're unsure. A weighted blanket is something else I keep at my office for people to try out and see if they like it and is the thing that people end up loving and going out and buying more than any of my other therapy hacks.

Natural Supports

Natural Supports are the people who love you just because you belong to them. Your family, your friends, teachers, coworkers, and others who go above and beyond their role in your life to support you getting better. Having people who love us just because they *do* is so, so, so important to getting better.

I understand that you may not have supportive people in your life, or your depression may be lying about which people care and want to see you better. If people offer support, lean into the discomfort of trusting that they mean it. If they ask to help, let them!

If you don't have people, it's time to find your people. Actively searching out friends when you already feel like shit is really difficult. Maybe start small and online, finding groups on social media where you can reach out to others with mental health issues, like the Facebook group for the Icarus Project. Maybe you can find a local support group. But honestly? It doesn't have to be a group of people who are also depressed . . . you can look for folks who also like movies, or reading, or hiking, or picking up and putting down heavy things. Getting out and exploring can be of great benefit to your healing, and bonus is you might meet some not-shitty humans.

It takes far more strength to accept help than reject. Be strong enough to allow others into your life and show you they care about you.

Peer Supports

There is a huge body of research that shows that peer-to-peer support partners (as the Substance Abuse and Mental Health Administration [SAMHSA] refers to them) are an enormous part of many people's wellness and recovery processes. This makes sense. Someone who has similar lived experience has a level of empathy, understanding, and compassion that other people don't. There are phenomenally caring treatment providers out there, but we often connect most to the people who have also traveled the same path we are on.

There are lots of names for this role in communities, including: recovery coach, sponsor, family partner, and systems navigator, to name a few. You will generally find these folks in recovery organizations (twelve-step or otherwise), community mental health clinics, and the like. There are also clinical professionals with lived experience who may share that experience as part of the work they do with people. If peer supports are available wherever you are seeking treatment, give it a try. Someone who has been in the same hole you are in is sometimes the best person to talk to about finding the way out, yanno?

Getting Back in Your Body and Mind

You've been reading through a lot of ideas for depression treatment that require the help of professionals. And, let's face it, that all costs money that you may not be able to spend right now. Even if you do have access to amazing treatment options, there is only so much that other people can do to help you. And all of what they can do works best when you are caring for yourself in the process. When it comes down to it, we have to be our own treatment providers.

The great irony is this: If you are fighting a mental health diagnosis, the very nature of the disease causes you to second-guess your own ability to care for yourself and heal. Depression is a hobgoblin of mind-fuckery.

So consider everything from here on down as reminders of how to take care of yourself. Reminders of the tools you have in your arsenal. Perspective on where you still have power, despite how fucked up your situation is or how much you have to fight like a damn warrior for every scrap of wellness you can muster.

Self-care

How much do humans suck at our own self-care? Almost all of us do . . . at least, until we crash and burn and are forcibly laid up because we literally made ourselves sick. Here is your reminder to stop that shit. Spend the energy you need to spend on getting better so you can better kick ass and chew bubblegum another day. Seriously, you are healing yourself.

I don't mean self-care in the Lifetime movie of the week sense of eating bonbons while taking a bubble bath (though there ain't nothing wrong with that). I mean showering and putting on clean PJs. Taking time to rest, read, journal, make art, go for a walk, listen to music. Be a little more gentle with yourself than you have in the past. This goes back to the depression funnel once again. We tend to leave out the things that seem the least important in the moment, even though they are the most important in the long run.

Mindful Movement/Exercise

Exercise is also very helpful for depression symptoms because it releases endorphins. And endorphins block our perception of pain and enhance positive feelings . . . both of which counterbalance the stress response. One study found that exercise is just as effective in reducing the symptoms of depression as therapy. Which means those super fit people who say they get a runner's high? Totally aren't lying. Freaks of nature, maybe, but telling the truth.

When I say exercise, I mean movement for fun. Movement for *play*. (Remember play?) No exercise as punishment because you had a damn cookie. This should be about choosing a form of play and movement that feels enjoyable for you.

I am not a fan of sweating and physical exertion in the name of health. But my doctor keeps telling me that reaching for a cookie does not count as a sit-up, so I gotta do *something*. I do enjoy swimming, walking, hiking, and yoga . . . they are way more relaxing and meditative for me than competitive team sports (but if that's your thing, go on with your weird-ass self!). Even better is when I go hiking with my bestie. We get exercise and get to talk shit about everyone we know in the process.

Find something that doesn't suck. It can be as intense or gentle as you want, but try stuff. Most places will offer a free class or free week, so check shit out. I had a client who fell in love with boxing by trying out a free class. It was great exercise *and* made her feel more empowered and in control of her experiences.

Some forms of exercise have demonstrated incredible benefit for treatment of depression beyond the general idea of *movement is good for us and endorphins are very helpful in reducing cortisol.* For example, significant research has been done on the use of yoga, and it has many long-term therapeutic effects in reducing physical pain *and* psychiatric symptoms. A specific method of yoga developed by David Emerson is even recognized in SAMHSA as an evidence-based practice.

Be A Shitty Meditator

I had someone tell me recently that they failed so hard at meditation that they haven't tried again in years. I SUCK SO

HARD AT MEDITATION, TOO! And that's why it's so fucking important and useful. The worse my sit, the more I learn about myself . . . for serious.

The point of meditation is not to be free of thoughts. The point is to recognize them, and then let that shit go. Yup, there's me making a grocery list. There's me singing the Arctic Monkeys in my head. There's me trying to figure out how much longer I have to meditate. There's me telling myself negative bullshit yet again.

Depression doesn't just lie, it diverts your attention. We don't usually catch our thoughts throughout the day. There is this running commentary that we have in our heads, a lot of which directly ties into the thinking errors I described earlier, that reinforces depression and our funnel slide. One study shows that mindfulness meditation decreases the chances of depression relapse better than antidepressant usage alone. How so? Another study on the topic watched biomarkers for stress and found them lower in people who use mindfulness meditation. So we can actually see the change in the body. That's pretty damn legit.

You can send me emails yelling at me that it totally isn't working. That you don't feel any better. And that you hate my stupid face for insisting that it's helpful. You may not feel any better, *but it still is helpful.* Because you know yourself better. You're paying attention to how shitty it really is. And that halts the funnel process. We are waking up to our own experiences when we meditate.

A Couple of Stupid-Simple Breathing Techniques

The 5-8-9 breathing pattern is an even simpler version of the breathing technique I published in my book *This Is Your Brain On Anxiety* (Microcosm Publishing, 2018). It creates a very similar effect because you are focusing on your breathing, and having your out-breath be longer than your in-breath, which calms your vagus nerve. (Reminder from earlier: This is your tenth cranial nerve that runs through your body and triggers your stress response head to fucking toe.) You simply breathe in for five, hold for eight, and blow out for nine. It's a breathing technique that Andrew Weil has written a lot about, but further back even than his work, we see it in hypnotherapy, neurolinguistic programming, and lots of places where mental health people were trying hard to work within the body instead of just the mind.

If 5-8-9 is more than you can handle at the moment, *try just breathing in and out, as slowly and deeply as you can, while counting to ten.* Like in-out "one," in-out "two," in-out "three," etc., etc. The breath cycles of 10 give you something different to focus on. Though what may happen is that you find yourself checking out and being up to 67 before you know it. And that's cool. Just remind yourself to start back at one again. Pause, regroup, and refocus.

Challenge Distorted Thinking Styles

There are a lot of aspects of depression that are completely out of our control. We can't trade in our biological relatives for a new set with better genes, right? (Though I know plenty of people who have tried.) But the research of depression recurrence shows that there are two certain things that we do have control over that can really impact our ability to manage depression and keep really bad depressive episodes from recurring, coming back as frequently, or being as severe as they could be.

These two things are what the National Institute of Health calls *cognitive vulnerability* and *neuroticism*.

Cognitive vulnerability means that you attribute negative shit happening (clinical term: "Stressful Life Events" [SLEs]) as somehow indicative of who you are as a person. Like you are the invoker of doom and bullshittery. People with cognitive vulnerability internalize and take responsibility for all the fucked-up things that happen to them, at least at some level.

When individuals with cognitive vulnerability are tracked over time, they show they are far more likely to suffer depression than other people. They start wiring themselves to expect SLEs. And since SLEs are going to fucking happen because fucked-up things happen, they mentally take on the bulk of responsibility for them, and over time they end up suffering clinical depression. This leads us back to Sapolsky's definition of depression as a learned helplessness response to stress. A clinical case of the fuck-its.

Neuroticism is more general. It means, essentially, that your default mode is negative. Not just that you internalize and take responsibility for the fucked-up things that happen, like mentioned above, but you are in a perpetual cycle of anger, anxiety, guilt, envy, etc. Everything in the world is colored by negativity. When we think of neuroticism, we typically think of any character played by Woody Allen in the history of ever, right? The people we think of as being neurotic we probably also consider exhausting pains in the ass. But in a clinical sense, people who lean toward the nervous and sensitive end of the scale are the people who are just more raw and vulnerable to the world. Neurotic people feel negativity at an entirely different level . . . and that shapes their worldview.

The big question is, then, can this shit be retrained?

Brains are really good at creating some fucked-up thinking patterns that continue to reinforce the groove of depressive behavior. People who use fancy therapeutic language refer to these patterns as *distorted thinking styles,* or the even more OG term *cognitive distortions.* When I was working at a partial hospitalization program, we had these distorted thinking styles listed and hung across the room. So during group process, anyone could refer to the list at any time and ask themselves, *"Which of these am I doing right now and how is that making this shit worse?"*

There is no hard and fast list of distorted thinking styles, and you may have seen different versions, but this one is based on one in the book *Thoughts and Feelings* (New Harbinger, 1981) so you will have a good starting point for paying attention to your own thinking patterns and the bullshit your brain is telling you. Once you start noticing it, then you can start counteracting the distorted thinking with what you would tell your best friend if you heard them trashing *themselves* the way you've been trashing *yourself.*

Filtering: Filtering happens when you pick out the one damn negative and don't look at anything else from the situation that is either positive or just neutral. I'm a professional trainer. You better believe I can teach 50 people and get 49 superlative reviews but perseverate on the one review that bitches about my potty mouth (fuck yeah!). With filtering, you are not including any context, because you are isolating what you are focusing on from the circumstances around it. Which means it becomes the entirety of your focus rather than just a percentage of your information.

Polarized Thinking: This is as black and white as a yin-yang symbol. Everything gets categorized as either good or bad. Then there is no middle ground. Which means if you aren't completely perfect, you must be a total failure.

Overgeneralization: This takes the polarized thinking to a whole new level. It's your brain taking one example of something that happened one time and deciding this is how things are going to roll forever. So if you failed at something, you are a failure and will continue to be so for time eternal. So why even bother trying?

Mind Reading: This is where you decide you know without a doubt why people are acting the way they act, saying what they say, and what they are thinking and feeling. Huge-ass leaps to conclusions. This is a lot of times based on projection. We know how we are thinking and feeling and presume that others would think and feel the same. Or we feel so shitty about ourselves that we presume others must have the same opinion of us. Someone didn't notice you? They totally must hate you, according to your brain. When, in reality, they are focused on why George R.R. Martin won't finish the damn *Game of Thrones* series or something that has nothing to do with you in that moment.

Personalization: So we all have the tendency to be the star of our own personal movie, right? And if we are the star, then everything going on around us must relate to us, right? So we are mind reading other people's intent as being about us instead of about them and presume that the ways they respond are due to something negative about us.

Control Fallacies: Fallacies of control fuck you up in either direction. If you feel totally controlled by everything around you, then you are totally a helpless victim. If you feel that you have to be in control of everything all the time, you feel responsible for *everything all of the time*. Neither is an empirical truth. Both extremes set you up for failure and exhaustion.

Fallacy of Fairness: I was pretty damn young when my mom told me, in essence, that life wasn't fair and I was exhausting myself by expecting it to be. If you are keeping a running fairness tally, you are going to be resentful to pissed off all the time.

Blaming: This is where you blame other people for the pain you feel and the problems you have. Now, people may fuck you over big time, but blaming them for everything that comes after means making them responsible for all of your choices and decisions for time eternal. That's far too much power to give over to some motherfucker. In reality, you have every right to be your own self-advocate, to make your own choices, and be responsible for your own decisions from here on out.

Shoulds: Dude. If people would just do what I say, the world would be a brilliant place. We all have a list of shoulds for the rest of the world. How people should behave. How they should treat us. When they don't follow those rules, we get all kinds of legit butthurt. Then we put on our judgey-pants about what they are doing, even when it's shit that we seriously don't even need to be worrying about.

Emotional Reasoning: This is where we presume that whatever emotion we are feeling is an indicator of something fucked up about ourselves. That if we feel something, it must be a fundamental truth. Like, if you feel guilty, you clearly *are* guilty. And while feelings are real, they don't always correspond to reality. If our thinking is distorted, then our emotional reactions go along for the ride and it becomes a mobius strip of mind-fuckery.

Fallacy of Change: This is the expectation that people should change to make life better for you. And that their changing is what is going to make you happy. Of course, you can always ask people to change. But the only person you can really control is your own damn self. And the choices you make will have far more of an impact on your happiness, because then you can take credit for successes.

Global Labeling: This is the shitty political extremist category of distorted thinking styles. It's all about stereotyping and one-dimensionality, as if knowing one thing allows you to know everything. It's what leads to prejudice, relationship issues, and the tendency to make snap judgments. For example, there is some dude in front of you in line at the store wearing a camo ball cap. Global labeling might mean automatically presuming he's a racist redneck just because you noticed the hat, but without having any other information about the dude.

Being Right: You feel like you have something to prove in every interaction . . . and what you need to prove is your inherent rightness. You are lawyer, judge, and jury . . . and you aren't hearing anything from the opposition. A different opinion or perspective? INADMISSABLE! Being right can really fuck up your ability to have caring and reciprocal relationships with others.

Heaven's Reward Fallacy: Heaven's Reward folks totally went to church with my grandma. They are the people who have some kind of sense that if they deny themselves and sacrifice constantly, they are somehow working toward some magical reward because there is a scorecard being kept. This isn't about people who are trying to be better human beings, because that's what we should all do, but a sort of false piety that speaks to paying into the system like it's an investment. Then, of course, when the reward doesn't come (because it doesn't have to be a literal storing of your treasures in heaven; maybe you are expecting public recognition of your moral superiority) it sucks pretty bad and creates serious bitterness. What was all the sacrifice for, then? If you aren't doing the right thing with your heart truly in it, you're just depleting yourself without real payoff.

Cognitive distortions are totally one of the ways the brain fucks us into a depressive episode. We are sliding down the funnel surrounded by our own negative thinking patterns. If we can capture and challenge these distortions, it can be incredibly helpful in keeping ourselves stable and functioning.

Count and Collect Spoons

Christine Miserandino created *spoon theory* to explain how living with illness and disability works to people who really struggle to understand it. You can check it out in its entirety at *ButYouDontLookSick.com*. Any metaphor will do. I've called it "fucks in my pocket" for decades, but "spoon theory" is G-rated and less likely to get you in trouble in certain crowds.

The idea is that we all have a certain number of (metaphorical) spoons. This isn't an equal number, however. If you have something going on, like pain, depression, etc. (whether this is a long-term thing or related to a specific current event), you have fewer spoons. People with more spoons don't understand why you are depleted all the time, right? If it took every spoon you had to get out of bed and shower and get dressed, then going for a walk can seem really overwhelming. You don't have an endless supply of spoons. And if you had a bunch of shit going on the day before, you may wake up in spoon deficit and really in need of recuperation.

Pay attention to the number of spoons you have. Look at what costs you spoons and what helps you build more spoons. This will help you make your best self-care decisions.

If you know that motivating yourself to go for a jog will cost 5 spoons, but you will feel better after to the tune of +10 spoons, then you want to set aside the five spoons you will need in order to do the thing that gives you a net gain of 5 extra spoons. Then you can use those motherfuckers to deal with your dentist appointment or going to the post office, right? It's a concrete cost-benefit analysis of your actions. And it's a good way of explaining where you are to people who are unlimited spoon-havers. You can say things like "*I have, like, two spoons left. I have enough for dinner or a movie, but not both right now. How about I join y'all after dinner for the movie? I'll sneak in the candy!*"

Spoon theory becomes a great shorthand metaphor for explaining your energy levels and planning around them. "So, how many spoons would that take?" is an idea I use with my clients a lot.

And when you find yourself really short on spoons, it can help you identify a narrowing of the funnel and give you the self-care reminder that you really needed.

Make a Crisis Plan

What's the point of a crisis plan? For me, the idea is that we all know what to do when we aren't well. It's super easy to name those things when we *are* well, but in the middle of a crisis (whether a depressive episode or a life-just-nut-punched-me episode), remembering to do these things is practically impossible unless you've been practicing doing so for years.

A crisis plan doesn't have to be fancy. The point is to have something down on paper that you can refer to when the bullshit hits the fan. These are the types of things you can personally commit to that can really help:

1) I will try to identify what is specifically upsetting me.
2) I will make a list of the things I can do in the here and now that are healthy for me, rather than causing me physical or emotional harm.
3) I will review the thoughts and conclusions I have made to see if I am being plagued by any distorted thinking styles. Even if my thinking is

pretty accurate, I will take a moment to evaluate how helpful it is to my state of mind.

4) I will do something that I enjoy and that helps me feel better for 30 minutes. Some of these activities can include: [THIS IS WHERE YOU WRITE DOWN SHIT YOU LIKE TO DO].
5) I will talk to someone I trust to be supportive of how I'm feeling. These people may include: [THIS IS WHERE YOU LIST YOUR PEEPS. PHONE NUMBERS, TOO, JUST IN CASE SOMEONE IS HELPING YOU CALL.].
6) I will repeat Steps 1–5 one more time. Give it a good chance to work.
7) If the thoughts continue and I am worried that I am not currently safe, I will contact my preferred crisis line. (There is zero shame in using a crisis line. You may just need to talk, to know that you aren't alone in the universe or in your experience. You may need more help and need someone to intervene on your behalf. A crisis line worker is trained to help you navigate the level of support you need in whatever your current situation. On your crisis plan, list the crisis lines that you feel will be most appropriate for you. There's a list

of them in the appendix if you need a starting point.) [NOW LIST THEM ON THE PLAN SO YOU AREN'T DIGGING AROUND FOR MY LIST LATER].

8) If I still feel that I am unsafe/unable to control myself, I will call 911 or go to the ER. My preferred ER is: [LIST THE ONE THAT SUCKS THE LEAST].
9) Contact info for other people that are important to helping me: [LIST YOUR EMERGENCY CONTACT, CASE MANAGER, PCP, PSYCHIATRIST, THERAPIST, SPONSOR, CLINIC WHERE YOU GET SERVICES, PASTOR, ETC.].

Mind the Funnel

So back to the funnel thing for a minute. When we start becoming mindful of when we start funneling down by noticing when our self-care lapses and our symptoms start getting worse, we are far more successful at heading off another depressive episode.

Start a simple list. What does being well look like for you? What are your early warning signs of depression? What are your seriously-down-in-the-funnel warning signs?

Once you have a good idea of what your personal funnel behaviors look like, create a new list with tools to battle each of them. For example, if you notice you aren't sleeping well, make a sleep hygiene plan for yourself. It might include no caffeine after a certain time, no TV in the bedroom, or a weighted blanket and essential oils to help you calm down.

Chances are you know which things work well, but it's hard to remember those things in the moment. Having an actual battle plan for each of them really helps.

And if you can't come up with ideas, it's a good thing to spend some time discussing with the clinicians you are receiving treatment from. Or your peer supports who make the kind of healthy choices you look up to.

6. CONCLUSION

Depression is another one of those tough-subject topics, I realize. It's hard to be cheerful about an illness that tends to eat people alive. But like everything else, I really believe that understanding the biochemical roots of the problem is enormously helpful in feeling less trapped and crazy. You are not defined by your depression. You are not weak, and you didn't do anything wrong. You didn't deserve this. You are not being punished. You hit the perfect storm of genetics + trigger and are now dodging and weaving while running your ass off toward the right end zone.

People struggling with depression (or any mental illness) are ANYTHING but crazy.

They are survivors, fighting back against brain chemistry that is entirely at odds with all the things that make life worth living.

Those of you who are living this? Who are saying, "*Fuck you, depression, you don't get to win today*"?

You are the bravest people I know.

Keep fighting.

7. APPENDICES

Resources for Battling Depression/Mood Disorders

Books

Hello Cruel World: 101 Alternatives to Suicide for Teens, Freaks, and Other Outlaws by Kate Bornstein

Alive with Vigor! Surviving Your Adventurous Lifestyle by Robert Earl Sutter III

How Not to Kill Yourself: A Survival Guide for Imaginative Pessimists by Set Sytes

Bluebird: Women and The New Psychology of Happiness by Ariel Gore

Maps to the Other Side: The Adventures of a Bipolar Cartographer by Sascha Altman DuBrul

The Price of Silence: A Mom's Perspective on Mental Illness by Liza Long

Furiously Happy by Jenny Lawson (but seriously, ANYTHING by Jenny Lawson)

Unfuck Your Brain by Faith G. Harper (SELF PROMOTION ALERT)

This Is Your Brain on Anxiety by Faith G. Harper (AGAIN?!)

Online resources

The Icarus Project: TheIcarusProject.net

The Icarus Project is designed to provide support and education outside of the traditional medical model for individuals with mental health diagnoses or symptoms. They are also huge advocates for social justice within the mental health system.

Bphope.com, Hope and Harmony for People with Bipolar: bphope.com

An online magazine and webforum that looks at the multiple ways bipolar disorder can affect individuals and their loved ones, exploring a whole host of treatment options and ideas.

Elephant Journal, It's About The Mindful Life: www.elephantjournal.com

A holistic wellness website with many contributors that speak with beautiful and brutal honesty about trauma and mental health issues.

Some Stats on Depression, Dysthymia, and Bipolar Disorder

The following statistics (collected from the National Survey on Drug Use and Health [NSDUH]) are provided to reiterate the truth that depression (and other mood disorders) are no joke. It is not a case of the blues. It's not being bummed out because you had a bad day at work. It is a very real disease that is costing our society in numerous ways, affecting every one of us, whether we suffer from depression or not. This information is provided to remind you of the following points: *You are not alone. You are not overreacting or being dramatic. This is serious shit that requires serious attention and serious treatment.*

- Major depression is one of the most common mental health disorders in the US, according to the National Institute of Mental Health (NIMH). The 2015 National Survey on Drug Use and

Health (NSDUH) estimated that 16.1 million Americans over the age of 18 had at least one major depressive episode in the past year. That's almost 7% of the population. And that's not a lifetime number. Just a snapshot of the year 2015.

- The NIMH estimates that another 1.5% of the population suffers from dysthymia and another 2.6% with a form of bipolar disorder.
- According to Mental Health America, up to 70% of individuals who complete suicide suffered from either depression or bipolar disorder.
- According to the National Institutes for Mental Health, up to 50% of people with bipolar disorder have attempted suicide at least once. This is due, at least in some part, to the struggle to get an appropriate diagnosis (like we talked about) and then get adequate treatment.
- The NIMH also admits that while nearly 83% of individuals with a diagnosis of bipolar disorder can be classified as severe ("mild" bipolar disorder is the exception, not the rule), less than 19% of people with bipolar disorder are receiving "*minimally adequate treatment.*"

- According to the World Health Organization, depression makes up the primary burden of all the mental and behavioral health related disability stats. Depression alone accounts for just over 8% of all the years lived with disability in the U.S.
- If it feels like this shit never goes away, that is a completely legit feeling. MDD has a really high recurrence rate, meaning if you have a depressive episode, the chances of having more is really fucking ridiculously high (as mentioned earlier in this book). According to the NIMH, if you have recovered from one episode, you have a 50/50 chance of having another and having it happen within 5 years of the first. If you've had two episodes in the past, you have an 80% likelihood of recurrence. Generally speaking, individuals with a history of depression will have about 5 separate depressive episodes in their lives. It's a vicious, nasty, soul-sucking cycle.

And let's talk about the economic burden of untreated depression. I include this information not to say, *"See what a horrible human being you are for struggling with depression?"* but instead to say, *"See! THIS is another reason we need to take depression seriously and get as serious about treating it as we do erectile dysfunction, FFS."* And also? This is great information to throw down with those asshats who don't much care about all those faceless people they don't know suffering day in and day out but do want a thriving economy to live and work in.

- The total economic burden of MDD is now estimated to be $210.5 billion per year.
- This is an increase of 21.5% between 2005 and 2010 (from $173.2 billion per year in 2005 to the $210.5 billion I mentioned above). To be fair, this is also a time period of a pretty severe economic crash. Definitely could be a contributing factor.
- About 45% of the costs are related to direct medical costs (medical services and pharmacy costs within employer-provided insurance).
- 5% are expenses related to suicide.
- Straight up half of the costs are related to *absenteeism* (missing work due to mood disorder symptomology) and what is called *presenteeism* (being at work, but having your productivity all jacked up due to symptoms. Presenteeism rose far more than absenteeism. The estimated workdays lost by people with depression because

of presenteeism was 32 days per person during this same 2005–2010 time span.

- Untreated depression causes a ton of other illnesses (like sleep problems and chronic pain), which has costs estimated at $4.70 for every dollar of direct MDD costs. Or, almost *five times as much* as what we are spending on the direct costs mentioned above.

Finding Treatment Providers

I swear, some days it feels like 99.44% of the battle is finding a provider who you can get in to see in a reasonable amount of time, really listens, is competent in treating you, and is a real partner in helping you get better.

You may be getting treatment in a community mental health setting, where you don't have much choice in providers, but if you do have choices, finding people who you really connect to and can work well with is important. So here are some things to consider asking when researching:

1) ***What is their license and/or certification? Where did they train? Who provides their practice oversight?***

Why These Questions Are Important: If the person is a licensed doctor or therapist or something, it's pretty obvious, but there are some service providers that may have exceptional training but not be licensed for any number of reasons. For example, less than half of the states in the US have a process for naturopathic doctors to become licensed. This doesn't mean that NDs working in states that do not provide licensing for them are suddenly poorly trained and incapable of doing their job. Asking these questions can help you weed out the sketch-ass people who have some certificate they bought online from the people who really know their shit.

2) ***What do they specialize in? What is their treatment approach? What is their training in your specialty areas?***

Why These Questions Are Important: I've seen so many people list specialties in which they had no real training or experience. That's setting everyone up for failure. It also could be that they specialize in a certain method of treatment that isn't what you are looking for even if they have TONS of experience and training. My treatment approach is pretty eclectic, so when people call me looking for a very structured treatment plan, I send them to someone who is great at that. I may have the training they are looking for, but not utilize that training in a way they would expect.

3) ***What is your experience in treating [insert your personal treatment needs]? Are you comfortable with working with [insert your issues]?***

Why These Questions Are Important: It gives you an idea of their ability to spitball when things go awry (and let's face it, most of us aren't gonna have every need met with a cookbook approach). It also may be that they don't have much experience with your issue but are willing to learn and are open to the challenge. And if they are someone you feel comfortable with and y'all work well as a team, that's OK. I have zero problem as a provider saying, "I have no idea, let me do some research and ask in my consulting group and look for some options for that issue."

4) ***These are other circumstances I have that could impact my care: [lay them all on the line]. Are you comfortable dealing with these issues?***

Why This Question Is Important: Some providers won't or can't work around all the stuff you have going on in your life, and it's better to know that right away. There may be stuff going on that makes working with a provider a deal-breaker. If you have other confounding medical issues. If you have other confounding mental health issues. If you are needing an assessment or diagnosis for other care or to receive benefits. If you are court involved and need the provider's testimony.

5) ***What do you charge? Are there additional fees for other services? How do you accept payment? Do you take insurance? Do you take HSA cards? Will you give me a superbill for reimbursement?***

Why These Questions Are Important: I try to give lots of information about my fees and options surrounding them. Not all providers (or their staff) think to do this, so it's really important that you ask. Especially if you are paying out of pocket and are having to budget and plan for your services . . . then, for example, you find out there is an additional fee for the diagnosis letter you were needing. Been there personally, and it really sucks.

6) Are you comfortable working with my other providers?

Why This Question Is Important: Because they may NEED to be in contact with your other providers. For example, if I am providing nutritional support to someone who is on prescription medications, I may need to chat with the prescribing docs about what I am wanting to treat and why. But sometimes it isn't even that complicated, and we just need to share info to better help our shared client. You want providers willing to do the legwork and pick up the phone and talk to the other people who are treating you.

Types of Prescription Antidepressants and Info on How They Work Best

All the info out there on antidepressants can be overwhelming and confusing. I've known tons of ridiculously smart people who had no idea what they were on, why they were on it, and what it was expected to do. That can lead to a helluva lot of problems.

So here is some basic info to get you started, with thanks to the Mayo Clinic and Aaron Sapp, MD, for the fact-checking and the reminders that what I see in my office as common isn't necessarily so common for the real world.

Medication Category	***How We Think They Work***	***What We Worry About***	***Examples***
Selective Serotonin Reuptake Inhibitors (SSRIs). Doctors often start by prescribing an SSRI. These medications generally cause fewer bothersome side effects and are less likely to cause problems at higher therapeutic doses than other types of antidepressants are.	SSRIs specifically affect serotonin levels by blocking the brain from recycling it (which is all reuptake means) during the neurotransmission process, leaving more serotonin available to combat depression.	The biggest issue I see complaints about is the sexual side effects. SSRIs can also affect sleep and increase irritability, agitation, and restlessness. Nausea and other common side effects are less likely to be long term, and may not be as bad with a different kind of SSRI since they all work a little differently.	• Fluoxetine (Prozac), • Paroxetine (Paxil, Pexeva) • Sertraline (Zoloft) • Citalopram (Celexa) • Escitalopram (Lexapro) • Vilazodone (Viibryd)

Medication Category	*How We Think They Work*	*What We Worry About*	*Examples*
Serotonin and Norepinephrine Reuptake Inhibitors (SNRIs). SNRIs help relieve both common and less common symptoms of depression. Sadness, irritability, and long term chronic nerve pain are all targeted by SNRIs.	SNRIs block the recycling (reuptake) of both serotonin and norepinephrine, leaving more of both available to combat depression.	Side effects of SNRIs are pretty similar to those of SSRIs.	• Duloxetine (Cymbalta) • Venlafaxine (Effexor XR) • Desvenlafaxine (Pristiq, Khedezla) • Levomilnacipran (Fetzima)

Medication Category	*How We Think They Work*	*What We Worry About*	*Examples*
Atypicals (Which only means the antidepressant catch-all category.)	This category includes several common antidepressants that don't fit well in one of the other categories but are thought to change levels of dopamine, serotonin and/or norepinephrine.	Side effects will all differ, obviously. Though it is important to note here that bupropion is the one antidepressant that does not have sexual side effects (though it does make you more susceptible to seizures, so it's not necessarily the magical wonder drug of the bunch).	• Bupropion (Wellbutrin/ Zyban, Forfivo XL, Aplenzin) • Mirtazapine (Remeron) • Nefazodone • Trazodone • Vortioxetine (Brintellix)

Medication Category	***How We Think They Work***	***What We Worry About***	***Examples***
Other Medications (Not Necessarily Antidepressants)	A prescriber may want to consider other medications in support of managing depression, including mood stabilizers, ADHD medications, anti-anxiety medications, medications for alertness, etc. Feel free to ask "Why this one specifically? What are you hoping to accomplish by having me take this?"	Varies	Varies

Medication Category	***How We Think They Work***	***What We Worry About***	***Examples***
RARELY PRESCRIBED Monoamine Oxidase Inhibitors (MAOIs).	These medications prevent at enzyme called monoamine oxidase from removing neurotransmitters associated with mood (norepinephrine, serotonin and dopamine from the brain) with the idea that they will then operate at proper levels, alleviating depression.	You have to maintain a really strict diet, because MAOIs can interact with certain foods, herbal supplements, and medications to the point of causing very high blood pressure. So these medications are typically used when other ones have failed.	• Isocarboxazid (Marplan) • Phenelzine (Nardil) • Selegiline (Emsam) • Tranylcypromine (Parnate)

Medication Category	*How We Think They Work*	*What We Worry About*	*Examples*
RARELY PRESCRIBED Tricyclic Antidepressants (TCAs). TCAs tend to cause more side effects than newer antidepressants, so they generally aren't prescribed unless you've tried other antidepressants first without improvement.	TCAs block the reabsorption (reuptake) of both serotonin and norepinephrine, leaving more of both available to combat depression.	TCAs block other chemical messengers in the body (not just serotonin and norepinephrine) so they are associated with more side effects than SSRIs and SRNIs, including tremors, excessive sleepiness, blood pressure issues, and weight issues.	• Imipramine (Tofranil) • Nortriptyline (Pamelor) • Amitriptyline • Doxepin • Desipramine (Norpramin) • Amoxapine • Protriptyline (Vivactil) • Trimipramine (Surmontil) • Maprotiline

What kinds of things do you and your prescriber need to consider when choosing an antidepressant?

1) Your particular symptoms. No-one has the exact same kind of depression, right? For example, if you sleep too much you might

respond better to an SSRI and if you sleep too little, you might do better with an atypical.

2) The medication side effects, your level of side effect sensitivity, and what you are willing to tolerate to get symptom relief.
3) What worked for your biological family members. If you have a parent or sibling that responded well to a certain antidepressant, then you are more likely to respond well to that one, too. And it can be a better starting place than just choosing one at random.
4) Cost and coverage. Some insurance companies insist on certain (cheaper) medications being tried first. Or, if your insurance is shitty (or you are paying out of pocket), it may make sense to try something cheaper before you have to start selling plasma just to keep yourself alive. Fortunately, some of the best antidepressants are pretty inexpensive, and many are on the four dollar list at a lot of pharmacies.
5) The other shit you are already taking. Medications have side effects enough in their own right; your prescriber is going to want to

avoid interactions with the other things you are already on.

6) Being pregnant or breastfeeding. Risks are generally pretty low, but certain antidepressants may be discouraged. There is a cost-benefit analysis conversation to have with your prescriber. Although Paxil is one of the ones that is more likely to be avoided for pregnant and breast-feeding moms, it was also the one that worked best for me when I had postpartum depression after having my daughter. So when I had my son, my doctor wanted me back on it, stating, “I’d far rather a little Paxil in your breastmilk than you being a crazy bitch again” (which also highlights why having a prescriber with whom you have a good relationship with is important).
7) Other health conditions. Some antidepressants can cause problems if you have other physical or mental health conditions. Though they aren’t always a bad thing . . . sometimes they can actually help other conditions; for example, Cymbalta can help with certain types of chronic pain, and

SSRIs can actually help manage premature ejaculation.

How can you optimize the chances of them working for you?

1) Take them as directed. You know, like consistently. And the prescribed dosage. And at the time of day you are supposed to take them, etc., etc. This can help their efficacy AND help manage side effects. So if your prescriber says to take Celexa at night, that is likely because it can make you drowsy, while Prozac and Zoloft tend to cause wakefulness, so you may be told to take them in the morning.
2) Be patient. This shit can take awhile. You may start noticing a difference in a couple weeks, but it can really take up to six weeks (or, ugh, longer) to get to the full treatment effect. And you may not be able to start off at the dose that's going to work best for you. You may need to start a low dose and level up gradually.
3) Communicate about the side effects. So many of my clients complain about the sexual side

effects of their meds to *me* but never told their prescriber. A good prescriber will say, "Hey, let me know if you notice [insert side effects here], because I totally have hacks to manage that or other ideas if we need to switch meds up," but they may not remember to do so. You gotta advocate for yourself.

4) Communicate if your other meds from other docs have changed. And communicate about what other supplements you are taking (herbal supplements in particular can interact badly with antidepressants).
5) Don't self-medicate unless your prescriber says its cool. This means alcohol, marijuana, anything else you might be using. Cuz this is seriously the time to be honest about your use. Or if you are afraid to admit to using something that isn't legal, at the very least ask your doctor, "Is there anything out there that people might be using that would be a really bad idea to use with what you are prescribing me?"
6) Don't stop taking them or fuck around with them without talking to your prescriber first.

Not because you need their permission to make decisions for yourself, but because some antidepressants can have really janky withdrawal effects that you may need help managing.

7) Do other stuff. Prescriptions aren't a be-all end-all. You will get better faster and stay better longer if you are getting therapy and doing other things for your general wellness like the exercise, and meditation, and better nutrition, and all that crunchy-granola shit that feels impossible then ends up being a lifesaver.

Herbs and Other Supplement Options

I have a postdoc in clinical nutrition, so I offer a lot of support in terms of diet and supplements for individuals looking for other complementary options for managing their depression. Whole food supplements are generally going to be safe to use in conjunction with prescription medications. Some herbal supplements can be contraindicated. That is, they end up making you sicker instead of better. The best rule of thumb is to ask your provider before adding something. I *always* communicate with prescribers about what I'm wanting to use and why and they *always* get the final word about anything I'm looking to add.

This is a basic list of adaptogenic herbs (herbs that help promote balance in the body, rather than force a particular upper or downer response) and nervines (herbs that have a calming effect specifically on the nervous system). This is not a be-all, end-all list. For example, I'm not trained in Chinese medicine, so if your practitioner is, you may see completely different combinations or names beyond my more limited knowledge in that area. My Chinese medicine practitioner has been a miracle worker on occasion, when I and my allopathic doc both failed, but even though I read up on whatever he gives me, it's still quite a mystery to me!

Adaptogens For Depression: Ashwagandha, Asian ginseng, holy basil, rhaponticum, rhodiola, and schisandra.

Adaptogens for Central Nervous System Support: Asian ginseng, rhaponticum, schisandra, shilajit, ashwagandha, cordyceps, and jiaogulan.

Adaptogens for Anxiety: Ashwagandha, jiaogulan, reishi, and schisandra.

Nervine Tonic Herbs (non-adaptogenic): Blue vervain, chamomile, fresh milky oat, hawthorn, linden, motherwort, passionflower, skullcap, lemon balm, St. John's wort, mimosa, lavender, and rosemary.

Other Supportive Supplements: Tryptophan (helps depression and sleeplessness), B12 (helps increase energy), GABA (helps manage anxiety), and trace minerals (helps manage irritability).

Crisis and Support Lines

I recently checked out all of these resources to make sure they were still operating and were legit resources. (I also fully expected to get a welfare check knock on the door by my local PD for doing so. Just sayin'.) Most of them are 24-hour lines, but not all of them, so it's a really good idea to have some backup numbers listed on your crisis plan, just in case. Also, these are all national numbers, so you will likely have different resources in the municipalities in which you live (the United Way 211 line can be a good starting point for finding your local info).

Suicide and Crisis Support Lines

Suicide Prevention Hotline	1-800-SUICIDE (784-2433)
Suicide Prevention Hotline En Espanol	1-888-628-9454
Suicide Prevention Hotline TTY Line (for individuals who are deaf or hard of hearing)	1-800-799-4889
TeenLine	1-800-TLC-TEEN
The Trevor Hotline (LGBT Crisis Line)	1-866-4-U-TREVOR
Trans Lifeline (Trans or GNC Crisis Line)	877-565-8860
United Way Helpline	211
ImAlive	Imalive.org
CrisisChat	CrisisChat.Org
Crisis Text Line	Text HOME to 741741
TeenLine Text	Text “TEEN” to 839863

Abuse and Violence Hotlines

National Sexual Assault Telephone Hotline	800-656-HOPE (4673)
National Domestic Violence Hotline	1-800-799-SAFE
National Domestic Violence Hotline En Espanol	1-800-942-6908
Stop It Now! (Sexual Abuse of Children Hotline)	1-888-PREVENT
Childhelp National Child Abuse Hotline	1-800-4-A-CHILD (1-800-422-4453)
National Center for Missing and Exploited Children	1-800-THE-LOST
National Domestic Violence Hotline	thehotline.org

Addictions Support Hotlines

Marijuana Anonymous	1-800-766-6779
24 Hour Cocaine Hotline	1-800-262-2463
SAMHSA's National Helpline (Treatment Referral Routing Service)	1-800-662-HELP (4357)
National Association for Children of Alcoholics	1-888-554-2627
National Problem Gambling Helpline Network	1-800-522-4700

LGBT Support Hotlines

Lesbian, Gay, Bisexual, and Transgender National Hotline	1-888-843-4564
Sage LGBT Elder Hotline	1-888-234-7243
LGBT National Youth Talkline	1-800-246-PRIDE
LGBT National Online Peer Support Chat	glbthotline.org/peer-chat.html

Homeless/Runaway Hotlines

Homeless/Runaway National Runaway Hotline	1-800-231-6946
Boys Town National Hotline	1-800-448-3000
National Runaway Safeline	1-800-RUNAWAY (786-2929)

Eating Disorder Hotlines

National Eating Disorders Association Helpline	1-800-931-2237
National Association of Anorexia Nervosa and Associated Disorders	1-847-831-3438

Non-Suicidal Self-Injury Hotline

Self-Injury Hotline	1-800-DON'T-CUT

Poison Control

Poison Helpline	1-800-222-1222

REFERENCES

Ackereley, R., et al. *"Positive Effects of A Weighted Blanket On Insomnia." https://www.jscimedcentral.com/SleepMedicine/sleepmedicine-2-1022.pdf*

"Antidepressants: Selecting one that's right for you." *Mayo Clinic, Mayo Foundation for Medical Education and Research*, 17 Nov. 2017, www.mayoclinic.org/diseases-conditions/depression/in-depth/antidepressants/art-20046273?pg=2

"Cyclothymia (Cyclothymic Disorder)." *WebMD*, WebMD, www.webmd.com/bipolar-disorder/guide/cyclothymia-cyclothymic-disorder#1

"Cyclothymia Personality, Symptoms, Causes, and Treatment." *PsyCom.net—Mental Health Treatment Resource Since 1986*, www.psycom.net/depression.central.cyclothymia.html

"Depression Treatment—Electrotherapy for Depression." *Alpha-Stim*, www.alpha-stim.com/alpha-stim-technology/depression/

Diagnostic and Statistical Manual of Mental Disorders: DSM-5. American Psychiatric Association, 2013.

Emerson, David. *Trauma-Sensitive Yoga in Therapy: Bringing the Body into Treatment*. W.W. Norton & Company, 2015.

Fox, Allison. "Does Meditation Really Help With Depression And Anxiety?" The Huffington Post, TheHuffingtonPost.com, 15 May 2017, www.huffingtonpost.com/entry/does-meditation-really-help-with-depression-and-anxiety_us_58cad68ee4b0ec9d29d9e4a7

Greenberg, P.E. et al. "The economic burden of adults with major depressive disorder in the United States (2005 and 2010)." The Journal of Clinical Psychiatry., U.S. National Library of Medicine, Feb. 2015, www.ncbi.nlm.nih.gov/pubmed/25742202

Hammond, D. Corydon, and Elsa Baehr. "Neurofeedback for the treatment of depression." *Introduction to Quantitative EEG and Neurofeedback*, 2009, pp. 295–313., doi:10.1016/b978-0-12-374534-7.00012-5.

Harvard Health. "What causes depression?" *Harvard Health*, www.health.harvard.edu/mind-and-mood/what-causes-depression

ICD 10: International Statistical Classification of Diseases and Related Health Problems. World Health Organization, 2009.

Lewinsohn, P. M., Joiner Jr., T. E., & Rohde, P. (2001). Evaluation of cognitive diathesis-stress models in predicting major depressive disorder in adolescents. *Journal of Abnormal Psychology, 110,* 203-215. doi10.1037/0021-843X.110.2.203

Lipton, Bruce H. *The Biology of Belief: Unleashing the Power of Consciousness, Matter & Miracles.* Hay House, Inc., 2016.

Lori. "Study: Herbal Products Omit Ingredients, Contain Fillers." *University of Guelph, 11 Oct. 2013,* www.uoguelph.ca/news/2013/10/study_herbal_pr.html

McKay, Matthew. *Thoughts & Feelings: Taking Control of Your Mood and Your Life.* New Harbinger, 2012.

Mindfulnext.org, mindfulnext.org/burnout-the-exhaustion-funnel/

"Monoamine oxidase inhibitors (MAOIs)." Mayo Clinic, Mayo Foundation for Medical Education and Research, 8 June 2016, www.mayoclinic.org/diseases-conditions/depression/in-depth/maois/art-20043992

"National Survey on Drug Use and Health." National Survey on Drug Use and Health (NSDUH), nsduhweb.rti.org/respweb/homepage.cfm

"Neurons, Synapses, Action Potentials, and Neurotransmission." *Neurons, Synapses, Action Potentials, and Neurotransmission—The Mind Project*, http://www.mind.ilstu.edu/curriculum/neurons_intro/neurons_intro.php

Penman, Danny. "Curing Depression with Mindfulness Meditation." Psychology Today, Sussex Publishers, 14 Oct. 2011, www.psychologytoday.com/blog/mindfulness-in-frantic-world/201110/curing-depression-mindfulness-meditation

"Postpartum Depression Facts." *National Institute of Mental Health, U.S. Department of Health and Human Services*, www.nimh.nih.gov/health/publications/postpartum-depression-facts/index.shtml

"Postpartum Depression (PPD) Center: Symptoms, Causes, Treatments, Medications, and Tests." *WebMD*, WebMD, www.

webmd.com/depression/postpartum-depression/default.htm

Porges, Stephen W. *The Polyvagal Theory: Neurophysiological Foundations of Emotions, Attachment, Communication, and Self-Regulation*. Norton, 2011.

Pulsipher, Charlie. "Natural vs. Synthetic Vitamins—What's the Difference?" Sunwarrior, Sunwarrior, 16 Aug. 2017, sunwarrior.com/healthhub/natural-vs-synthetic-vitamins

"Reiki & the Urban Zen Integrative Therapy Program @ UCLA." *Healing Your Broken Heart, Body, Mind and Soul!*, 18 Oct. 2017, reikienergyhealingla.org/reiki-scientific-research-clinica-studies/reiki-the-urban-zen-integrative-therapy-program-ucla/

Roth, Lauren. "Upper Cervical Chiropractic Care as a Complementary Strategy for Depression and Anxiety: A Prospective Case Series Analysis." *Journal of Upper Cervical Chiropractic Research* (June 20, 2013): 49-59.

Sapolsky, Robert M. *Why Zebras Don't Get Ulcers: An Updated Guide to Stress, Stress-Related Diseases, and Coping*. W.H. Freeman, 2001.

"Selective serotonin reuptake inhibitors (SSRIs)." *Mayo Clinic, Mayo Foundation for Medical Education and Research,* 24 June 2016, www.mayoclinic.org/diseases-conditions/depression/in-depth/ssris/art-20044825

"Serotonin and norepinephrine reuptake inhibitors (SNRIs)." *Mayo Clinic, Mayo Foundation for Medical Education and Research,* 21 June 2016, https://www.mayoclinic.org/diseases-conditions/depression/in-depth/antidepressants/art-20044970

"The Spoon Theory written by Christine Miserandino." *But You Don't Look Sick? support for those with invisible illness or chronic illness*, 26 Apr. 2013, butyoudontlooksick.com/articles/written-by-christine/the-spoon-theory/

"Tricyclic antidepressants (TCAs)." *Mayo Clinic, Mayo Foundation for Medical Education and Research,* 28 June 2016, www.mayoclinic.org/diseases-conditions/depression/in-depth/antidepressants/art-20046983

Turek, Fred. "Faculty of 1000 evaluation for Induction of photosensitivity by heterologous expression of melanopsin." *F1000—Post-Publication peer review of the biomedical literature*, 2005, doi:10.3410/f.1023715.297788

Weintraub, Amy. *Yoga for Depression: A Compassionate Guide to Relieving Suffering through Yoga.* Broadway Books, 2004.

Williams, Mark G., et al. *The Mindful Way through Depression: Freeing Yourself from Chronic Unhappiness.* Guilford Press, 2007.

Winston, David, and Steven Maimes. *Adaptogens: Herbs for Strength, Stamina, and Stress Relief.* Healing Arts Press, 2007.

Wolynn, Mark. *It Didn't Start with You: How Inherited Family Trauma Shapes Who We Are and How to End the Cycle.* Penguin Books, 2017.

I have the whole 5 MINUTE THERAPY set, but I have to keep buying them again because my friends need them too!
GRIEF
DEPRESSION
RELATIONSHIPPING
ANGER
DATING
ADDICTION
DEFRIENDING
ADULTING
COPING SKILLS
ANXIETY
How Not to Kill Yourself

UNF*CK YOUR BRAIN

GETTING OVER ANXIETY, DEPRESSION, ANGER, FREAK-OUTS, AND TRIGGERS...WITH SCIENCE!

FAITH G. HARPER, PHD, LPC-S, ACS